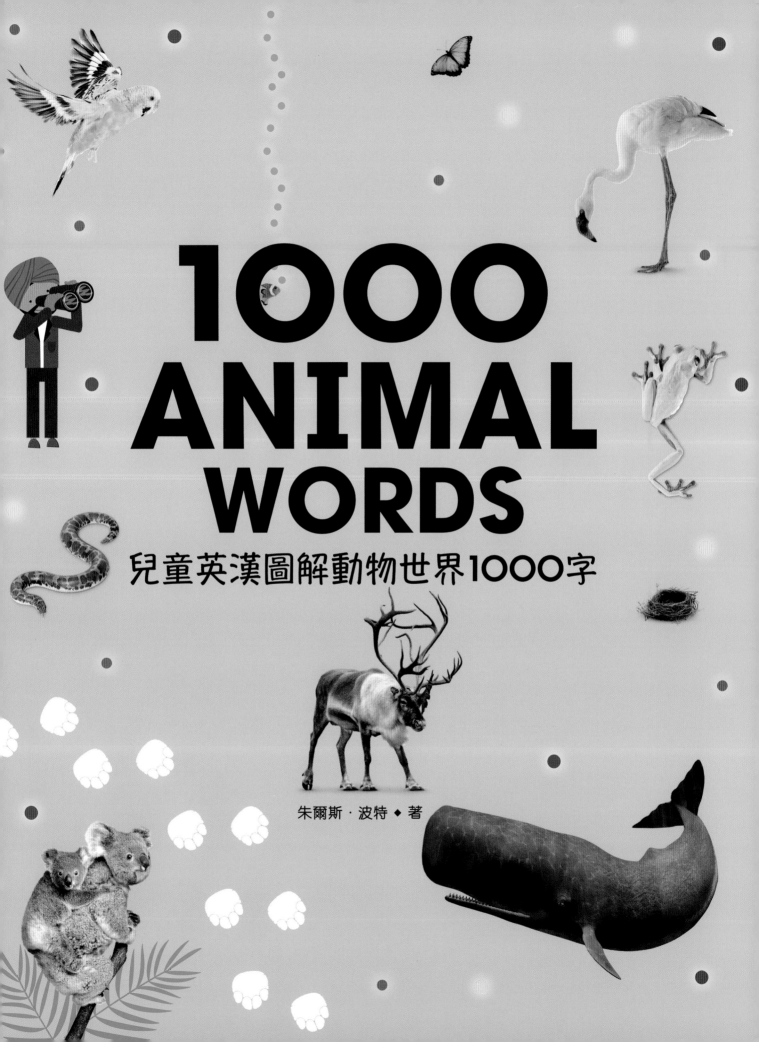

1000 ANIMAL WORDS

兒童英漢圖解動物世界1000字

朱爾斯‧波特 ◆ 著

1000 ANIMAL WORDS

兒童英漢圖解動物世界 1000 字

作者：朱爾斯・波特（Jules Pottle）

翻譯：亦其

責任編輯：陳奕祺

美術設計：張思婷

出版：新雅文化事業有限公司

香港英皇道499號北角工業大廈18樓

電話：（852）2138 7998

傳真：（852）2597 4003

網址：http://www.sunya.com.hk

電郵：marketing@sunya.com.hk

發行：香港聯合書刊物流有限公司

香港荃灣德士古道220-248號荃灣工業中心16樓

電話：（852）2150 2100　傳真：（852）2407 3062

電郵：info@suplogistics.com.hk

版次：二〇二三年四月初版

ISBN: 978-962-08-8147-3

Original Title: *1000 Animal Words: Build Animal Vocabulary and Literacy Skills*

Copyright © Dorling Kindersley Limited, 2023

A Penguin Random House Company

Traditional Chinese Edition © 2023 Sun Ya Publications (HK) Ltd.

18/F, North Point Industrial Building, 499 King's Road, Hong Kong

Published in Hong Kong SAR, China

Printed in China

For the curious
www.dk.com

混合產品
紙張 |
支持負責任的林業
FSC® C018179

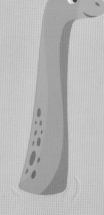

1OOO ANIMAL WORDS

兒童英漢圖解動物世界1OOO字

朱爾斯·波特 ◆ 著

新雅文化事業有限公司
www.sunya.com.hk

給爸爸媽媽的話

珍惜動物從小開始

　　什麼是動物？如果你問孩子，他們可能會說出寵物或動物園裏的動物名稱。這類動物許多都是毛茸茸或色彩斑斕的，或像我們一樣有大眼睛。這些動物都為我們熟悉，然而，牠們不是唯一重要的動物。

　　孩子在日常生活中遇到的動物，例如蜘蛛和昆蟲，有可能被認為是害蟲。一些動物則因為太細小或暗淡無光，無法引起孩子的注意。但是，這些看似微不足道，在日常生活隨處可見的動物，對維持生態系統平衡，卻與熊、大象或袋鼠等大型、顯而易見的物種，一樣重要。

　　動物世界的多樣性非常有趣，由雨林中色彩鮮豔的鳥類，到竹節蟲這類偽裝大師都有。在維繫地球的微妙平衡上，每個物種都扮演牠們的角色，所以我們要學習欣賞每個物種，甚至每一個人的價值，而欣賞需從知識和理解出發。

　　在這個世代，我們要比以往更尊重動物，學習保護牠們。這本書將動物學中各種有趣的事情，獻給那些閱讀這本書的年輕動物愛好者，培養他們對動物的好奇心，珍惜動物。請與孩子一起探索本書的每一頁，讓它啟發孩子思考動物與人類的關係，也讓我們教育下一代好好保護地球上的不同物種。

朱爾斯・波特 (Jules Pottle)
資深小學科學科顧問、教師、導師及作家

Contents 目錄

Classifications 動物分類

Animals are classified as vertebrates or invertebrates. Within those two groups there are smaller groups. Here are some of them.

動物分為脊椎動物和無脊椎動物。在這兩個大類中，會分支出細小的類別，以下便是其中一些類別。

luna moth
月亮蛾

harvest mite
秋蟎

peacock spider
孔雀蜘蛛

Insects
昆蟲

Arachnids
蛛形綱動物

leaf insect
葉竹節蟲

Arthropods
節肢動物

signal crayfish
信號小龍蝦

barrel sponge
桶狀海綿

acorn barnacle
藤壺

European rhinoceros beetle
犀角金龜

Crustaceans
甲殼動物

Sponges
海綿

peanut worm
星蟲

banana slug
香蕉蛞蝓

Worms
蟲

Halloween hermit crab
溝紋纖毛寄居蟹

Molluscs
軟體動物

coconut octopus
椰子章魚

horsehair worm
鐵線蟲

Atlantic surf clams
大西洋衝浪蛤蜊

Invertebrates 無脊椎動物

cockatiel
雞尾鸚鵡

Birds
鳥類

ring-billed gull
環嘴鷗

greater
roadrunner
走鵑

Somali giraffe
網紋長頸鹿

quokka
短尾矮袋鼠

great crested newt
大冠蠑螈

golden
toad
金蟾蜍

Mammals
哺乳動物

Amphibians
兩棲動物

grey whale
灰鯨

fire salamander
火蠑螈

koi
錦鯉

Fish
魚類

leopard gecko
豹紋守宮

Reptiles
爬蟲動物

hawksbill
sea turtle
玳瑁

stingray
魔鬼魚

Vertebrates 脊椎動物

Invertebrates 無脊椎動物

Most animal species are invertebrates with no internal bony skeleton.
大部分動物都屬於體內沒有骨骼的無脊椎動物。

Arthropods 節肢動物

Insects
昆蟲

ant
螞蟻

six legs
6隻腳

wasp
黃蜂

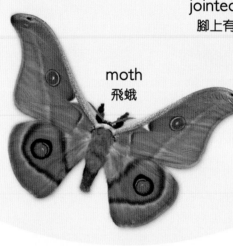
moth
飛蛾

jointed legs
腳上有關節

shield bug
nymph
盾蝽若蟲

Arachnid
蛛形綱動物

spider
蜘蛛

eight legs
8隻腳

Which has more legs:
an arachnid or an insect?
猜猜是蛛形綱動物還是昆蟲的腳多？

tick
硬蜱

Crustaceans
甲殼動物

woodlouse
鼠婦

shrimp
蝦

crab
蟹

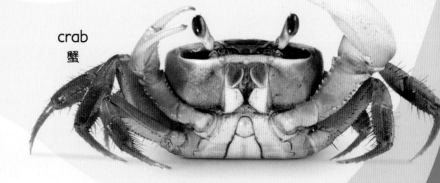

Myriapods
多足類動物

millipede
千足蟲

All invertebrates share these features.
所有無脊椎動物都
具有這些特徵。

no
bony skeleton
沒有骨骼

cold-blooded
變溫動物

no spine
沒有脊椎

lay eggs
產卵

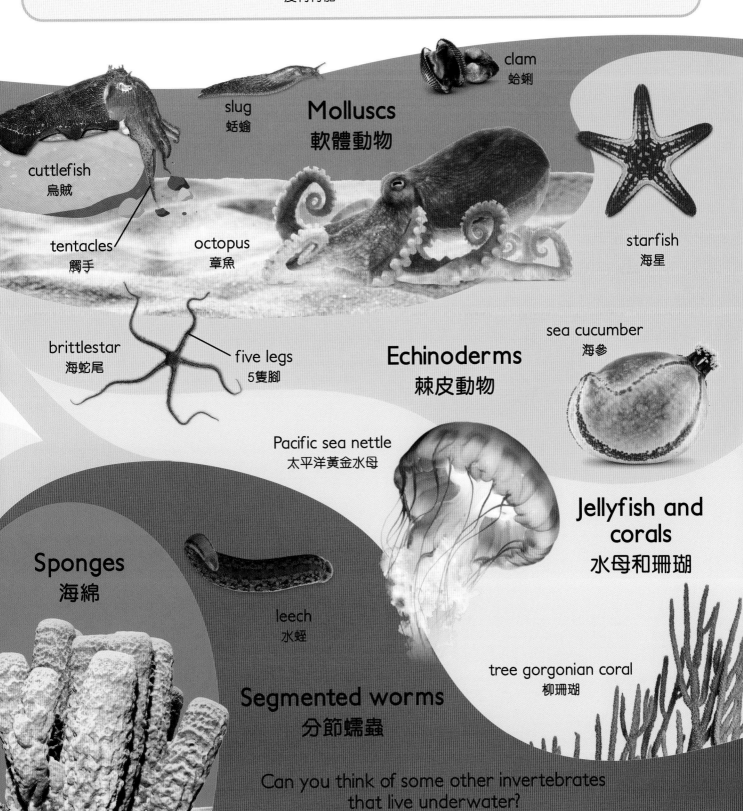

slug
蛞蝓

clam
蛤蜊

Molluscs
軟體動物

cuttlefish
烏賊

tentacles
觸手

octopus
章魚

starfish
海星

brittlestar
海蛇尾

five legs
5隻腳

Echinoderms
棘皮動物

sea cucumber
海參

Pacific sea nettle
太平洋黃金水母

Jellyfish and corals
水母和珊瑚

Sponges
海綿

leech
水蛭

tree gorgonian coral
柳珊瑚

Segmented worms
分節蠕蟲

Can you think of some other invertebrates that live underwater?
你能想到其他生活在水中的無脊椎動物嗎？

Insects 昆蟲

There are more species of insects than any other type of animal on Earth.

昆蟲的種類比地球上任何動物的種類都要多。

dung beetle
(scarab)
糞金龜(蜣螂)

Beetles
甲蟲

rose chafer
金花金龜

rhinoceros
beetle
犀角金龜

head
頭部

thorax
胸部

abdomen
腹部

Goliath
beetle
大角金龜

bumble bee
熊蜂

common
wasp
普通黃胡蜂

woolly apple
aphid
蘋果綿蚜

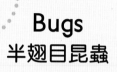

common
pond skater
小黽蝽

Bugs
半翅目昆蟲

greengrocer cicada
澳洲綠蟬

shield bug
盾蝽

monarch butterfly
帝王斑蝶

Bees, wasps, and ants
蜜蜂、黃蜂和螞蟻

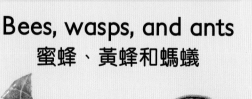

leafcutter
ant
切葉蟻

cocoon
繭

Atlas moth
皇蛾

Butterflies
and moths
蝴蝶和飛蛾

egg
卵

caterpillar
毛蟲

common
bluetail damselfly
褐斑異痣蟌

emperor dragonfly
豆娘

desert
locust
沙漠蝗蟲

Crickets, locusts,
and grasshoppers
蟋蟀、蝗蟲和蚱蜢

Dragonflies and
damselflies
蜻蜓

nymph
若蟲

common field grasshopper
常見田野蚱蜢

great green cricket
綠叢螽斯（螽：粵音宗）

Cockroaches
蟑螂

Termites
白蟻羣

Fleas
蚤

American
cockroach
美洲蟑螂

globe skimmer
dragonfly
黃蜻

hissing
cockroach
馬達加斯加蟑螂

termite
白蟻

dog flea
狗蚤

Earwigs
蠼螋
（粵音渠搜）

common mayfly
常見蜉蝣

European earwig
歐洲蠼螋

Mayflies
蜉蝣

Mediterranean
praying mantis
屬模虹螳

leaf insect
葉竹節蟲

Stick insects and
leaf insects
竹節蟲及葉竹節蟲

Praying
mantises
螳螂

Vietnamese
stick insect
越南竹節蟲

Birds 鳥類

Birds have beaks and feathers. They lay eggs with hard shells. Birds that fly have hollow bones to make them lighter.

鳥類有喙和羽毛。牠們會產下有硬殼的蛋。會飛的鳥有空心骨頭，使牠們體型更輕，便於飛行。

Seabirds
海鳥

masked booby
藍臉鰹鳥

wandering albatross
漂泊信天翁

Atlantic puffin
北極海鸚

common gull
海鷗

Birds of prey
猛禽

osprey
魚鷹

sparrowhawk
雀鷹

barn owl
倉鴞

Water birds
水鳥

Which birds are native to where you live?

哪些鳥是你居住地的本土物種？

mute swan
疣鼻天鵝

mallard duck
綠頭鴨

Canada goose
加拿大雁

Wading birds
涉禽

greater flamingo
大紅鸛

yellow-billed stork
黃嘴鸛鸛

All birds share these features.
所有鳥類都具有這些特徵。

 bony skeleton 有骨骼

warm-blooded 恆溫動物

 hard-shelled eggs 產下硬殼蛋

feathers 羽毛

 scaly legs 腳上有鱗片

Flightless birds
不能飛的鳥

emu
鴯鶓
（響音而苗）

common ostrich
鴕鳥

kiwi
鷸鴕
（奇異鳥）

bald eagle
白頭海鵰

Penguins
企鵝

emperor penguin
皇帝企鵝

gentoo penguin
巴布亞企鵝

common swift
普通樓燕

Domestic birds
家禽

 chicken
雞

Small birds
小型鳥類

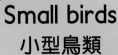 common wood pigeon
斑尾林鴿

 blue tit
藍山雀

 house sparrow
家麻雀

Parrots
鸚鵡

galah
粉紅鳳頭鸚鵡

scarlet macaw
緋紅金剛鸚鵡

 ruby-throated hummingbird
紅玉喉北蜂鳥

 European robin
知更鳥

Amphibians 兩棲動物

Most amphibians live on land for much of the year. They must return to the water to breed because their eggs have no shells.

大部分兩棲動物，全年中多數時間都生活在陸地上。牠們必須回到水中繁殖，因為牠們產下的卵沒有殼。

Frogs and toads
青蛙和蟾蜍

Salamanders
蠑螈

American toad tadpole
美國蟾蜍蝌蚪

American toad
美國蟾蜍

natterjack toad
黃條背蟾蜍

American bullfrog tadpole
美國牛蛙蝌蚪

American bullfrog
美國牛蛙

croak
呱呱

some oxygen absorbed through skin
部分氧氣透過皮膚吸收

cane toad
海蟾蜍

common frog
林蛙

common frog tadpole
林蛙蝌蚪

All amphibians
share these
features.
所有兩棲動物都
具有這些特徵。

bony
skeleton
有骨骼

cold-blooded
變溫動物

jelly eggs
水中產卵

adult
has lungs
成年有肺部

juvenile
has gills
幼體有鰓

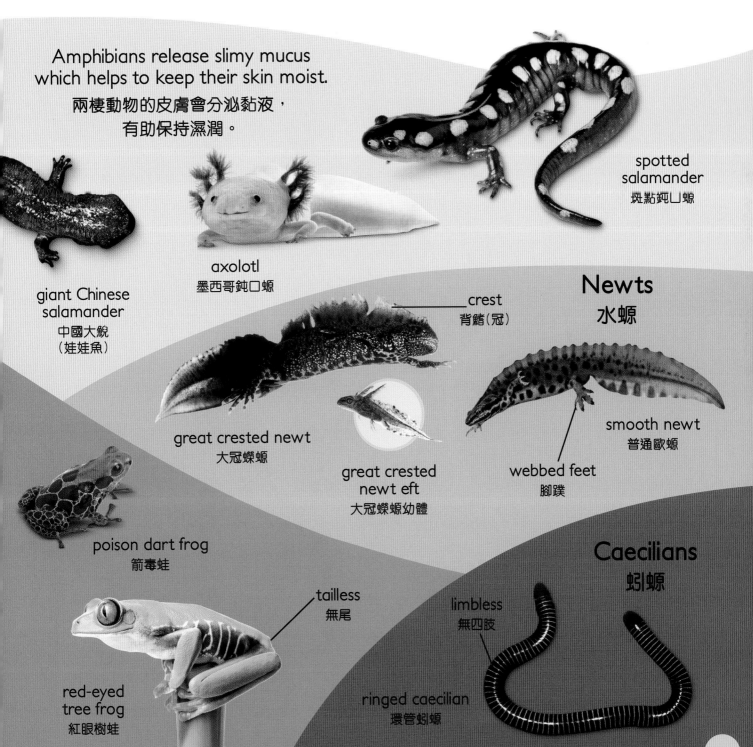

Amphibians release slimy mucus
which helps to keep their skin moist.
兩棲動物的皮膚會分泌黏液，
有助保持濕潤。

spotted
salamander
斑點鈍口螈

axolotl
墨西哥鈍口螈

giant Chinese
salamander
中國大鯢
（娃娃魚）

crest
背鰭（冠）

**Newts
水螈**

great crested newt
大冠蠑螈

great crested
newt eft
大冠蠑螈幼體

webbed feet
腳蹼

smooth newt
普通歐螈

poison dart frog
箭毒蛙

tailless
無尾

limbless
無四肢

**Caecilians
蚓螈**

red-eyed
tree frog
紅眼樹蛙

ringed caecilian
環管蚓螈

Mammals 哺乳動物

Mammals give birth to live young which feed on their mother's milk.

哺乳動物誕下幼崽，並以母乳餵哺。

Most mammals share these features.
大部分哺乳動物都具有這些特徵。

brown-throated sloth
褐喉樹懶

Bactrian two-humped camel
雙峯駱駝

mongoose lemur
獴美狐猴

bonnet macaque
冠毛獼猴

mouse
老鼠

Burchell's zebra
布氏斑馬

llama 大羊駝

African elephant
非洲象

Some mammals, such as whales and dolphins, live underwater.
一些哺乳動物，如鯨和海豚，生活在水中。

North American river otter
北美水獺

sperm whale
抹香鯨

bottlenose dolphin
寬吻海豚

bony skeleton
有骨骼

warm-blooded
恆溫動物

pregnant
胎生

live young
幼崽

mother's milk
哺乳

most have fur
大部分有毛

pangolin
穿山甲

bat
蝙蝠

ermine
白鼬

red panda
小熊貓

capybara
水豚

sun bear
馬來熊

warthog
疣豬

brown hare
歐洲野兔

ringed seal
環斑海豹

Virginia opossum
北美負鼠

Marsupials
有袋動物

wombat
袋熊

wallaby
小袋鼠

North America
北美洲

South America
南美洲

New Guinea
新幾內亞

koala
樹熊

Australia
澳洲

Tasmania
塔斯曼尼亞州

undeveloped young
幼獸

Monotremes
卵生哺乳類

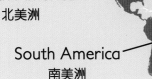

lay eggs
產卵

New Guinea
新幾內亞

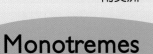

Australia
澳洲

echidna
澳洲針鼴
（鼴：粵音演）

duck-billed platypus
鴨嘴獸

17

Primates 靈長類

Humans belong to the primate family. Our brains are highly complex, just like other primates.

人類屬於靈長類動物。我們的大腦與其他靈長類動物一樣非常複雜。

howler monkey
吼猴

golden lion tamarin
金獅面狨

emperor tamarin
皇狨猴

grasp
抓着

woolly spider monkey
絨毛蛛猴

spider monkey
蜘蛛猴

use tools
利用工具

Monkey tails can grip branches, like an extra hand.
猴子可以用尾巴夾住樹枝，就像牠第三隻手。

swing through trees
在樹木間盪來盪去

squirrel monkey
松鼠猴

capuchin monkey
捲尾猴

woolly monkey
絨毛猴

titi
伶猴

gibbon
長臂猿

aye-aye
指猴

ring-tailed lemur
環尾狐猴

vervet monkey
青腹綠猴

olive baboon
東非狒狒

loris
蜂猴

Barbary macaque
巴巴里獼猴

Abyssinian
black-and-white
colobus (guereza)
阿比西尼亞黑白疣猴

rhesus monkey
普通獼猴

talk
會說話

human
人類

chimpanzee
黑猩猩

bonobo
倭黑猩猩

gorilla
大猩猩

no tail
無尾

proboscis
monkey
長鼻猴

orangutan
紅毛猩猩

19

Fish 魚類

There are two types of fish. One type has a skeleton made of bone. The other has a skeleton made of cartilage, which is more flexible.

魚可以分為兩大類，一類具有由骨頭製成的骨架；另一類具有由軟骨製成的骨架，活動更靈活。

All fish share these features.
所有魚類都具有這些特徵。

cold-blooded
變溫動物

slender seahorse
吻海馬

freshwater angelfish
大神仙魚

batfish
蝙蝠魚

Atlantic salmon
大西洋三文魚

coelacanth
腔棘魚

Atlantic bluefin tuna
北方藍鰭金槍魚

blue regal tang
擬刺尾鯛

puffer fish
河豚

clownfish
小丑魚

electric eel
電鰻

European plaice
歐洲鰈

angler fish
鮟鱇魚

blue catfish
藍鯰魚

Bony fish 硬骨魚

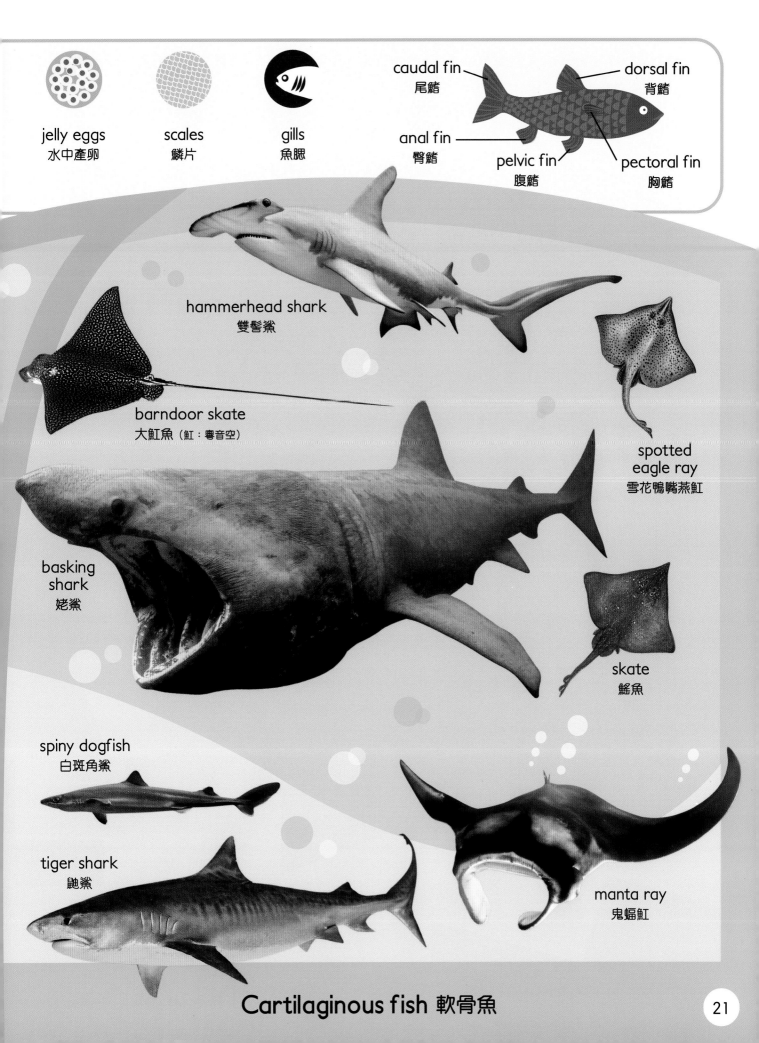

jelly eggs
水中產卵

scales
鱗片

gills
魚腮

caudal fin
尾鰭

dorsal fin
背鰭

anal fin
臀鰭

pelvic fin
腹鰭

pectoral fin
胸鰭

hammerhead shark
雙髻鯊

barndoor skate
大魟魚（魟：粵音空）

spotted
eagle ray
雪花鴨嘴燕魟

basking
shark
姥鯊

skate
鰩魚

spiny dogfish
白斑角鯊

tiger shark
鼬鯊

manta ray
鬼蝠魟

Cartilaginous fish 軟骨魚

Reptiles 爬行動物

Reptiles are cold-blooded: they cannot make their own body heat. Most reptiles must bask in the sun to warm up before they can be active.

爬行動物是變溫動物，牠們不能讓身體自行發熱。
大多數爬行動物必須先曬太陽取暖，才能活動。

Lizards
蜥蜴

green iguana
綠鬣蜥

panther chameleon
豹變色龍

Komodo dragon
科莫多龍

leopard gecko
豹紋守宮

diamondback terrapin
鑽紋龜

Reptiles have been around for over 300 million years.
爬行動物已存在了超過3億年。

Tortoises and turtles
陸龜和海龜

crawl
爬行

shell (carapace)
龜甲

scute 盾板

Blanding's turtle
布氏擬龜

red-footed tortoise
紅腿象龜

horny beak
角質喙

Galapagos giant tortoise
加拉巴哥象龜

legs
腳

claws
爪子

All reptiles share these features.
所有爬行動物都具有這些特徵。

bony skeleton
沒有骨骼

cold-blooded
變溫動物

leathery eggs
革質蛋

scales
鱗片

black mamba
黑曼巴蛇

Snakes
蛇

boa constrictor
紅尾蚺

shed skin
蛻皮

slither
滑行

fangs
毒牙

forked tongue
分叉舌頭

Tuataras
大蜥蜴

tuatara
大蜥蜴

only in New Zealand
僅存於新西蘭

tail ridge
尾脊

Crocodiles and alligators
鱷魚和短吻鱷

belly crawl
腹部貼地爬行

Nile crocodile
尼羅鱷

American alligator
美國短吻鱷

webbed feet 腳蹼

23

Our pets 我們的寵物

Humans enjoy the company of animals. We keep many different types of animals as pets.

人類喜歡有動物作伴。我們飼養了許多不同類型的動物作為寵物。

terrapin
水龜

aquarium
水族箱

tropical fish
熱帶魚

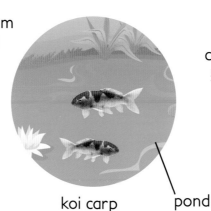

koi carp
錦鯉

pond
池塘

puppy
小狗

collar
頸圈

bed
牀

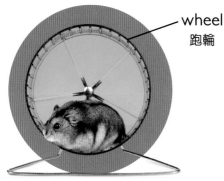

wheel
跑輪

hamster
倉鼠

gerbil
沙鼠

parrot
鸚鵡

ferret
雪貂

* 部分地區禁止飼養雪貂。

pony
小馬

rat
大鼠

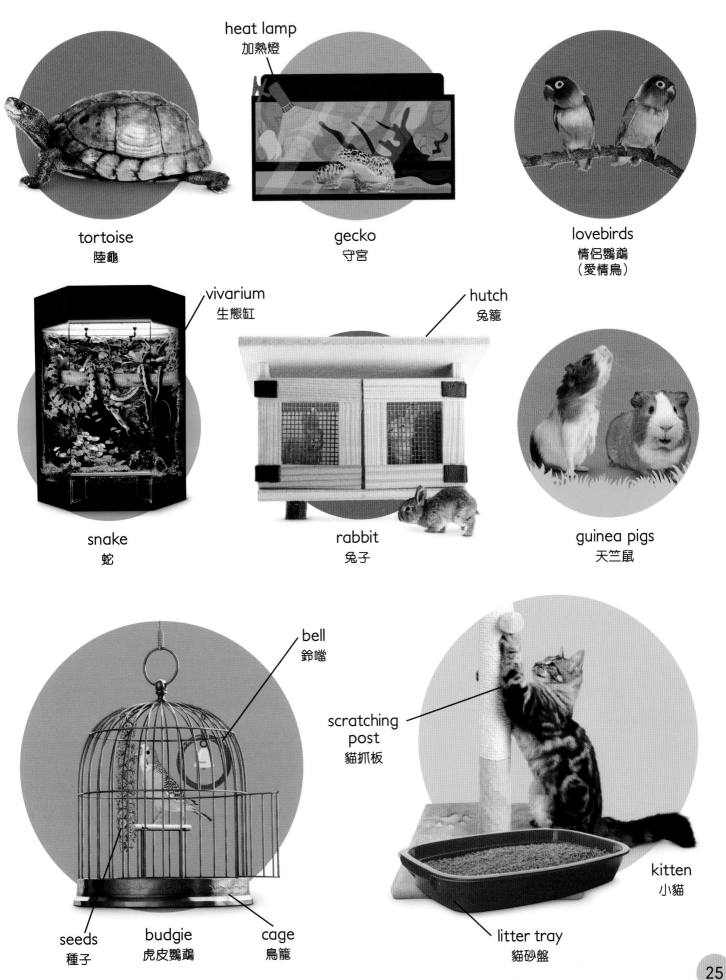

tortoise
陸龜

heat lamp
加熱燈

gecko
守宮

lovebirds
情侶鸚鵡
（愛情鳥）

vivarium
生態缸

snake
蛇

hutch
兔籠

rabbit
兔子

guinea pigs
天竺鼠

bell
鈴噹

scratching
post
貓抓板

seeds
種子

budgie
虎皮鸚鵡

cage
鳥籠

litter tray
貓砂盤

kitten
小貓

Domestic dogs 家犬

Humans have kept dogs for thousands of years. Many breeds are working dogs. Others are bred to be good company.

人類養狗已有數千年歷史。許多品種都是工作犬，其他則被培養成人類的好朋友。

Do you have a favourite dog breed?
你有最喜歡的狗品種嗎？

以玩具球圖示對比狗隻體型，就能知道狗隻的實際大小。

Hounds
獵犬

Beagle
小獵犬

Basset Hound
巴吉度獵犬

Whippet
惠比特犬

Irish Wolfhound
愛爾蘭獵狼犬

Greyhound
格雷伊獵犬

Retrievers
尋回犬

Dachshund
臘腸狗

Irish Setter
愛爾蘭雪達犬

Norwegian Elkhound
挪威獵糜犬

Cocker Spaniel
美國曲架

Labrador Retriever
拉布拉多

Golden Retriever
金毛尋回犬

Herding dogs
牧羊犬

German Shepherd
德國牧羊犬

Newfoundland
紐芬蘭犬

Old English Sheepdog
英國古代牧羊犬

Boxer
拳師犬

Samoyed
西摩犬

Siberian Husky
西伯利亞雪橇犬

Border Collie
邊境牧羊犬

Doberman Pinscher
都柏文

Utility dogs
工作犬

Yorkshire Terrier
約瑟爹利

Jack Russell
積羅素犬

Toy dogs
玩賞犬

Chihuahua
芝娃娃

French Bulldog
法國鬥牛犬

Toy Poodle
玩具貴婦狗

Lap dogs
哈巴狗

Cockapoo
曲架頗犬

Pug
八哥

Cavalier King
Charles Spaniel
騎士查理王獵犬

Shih Tzu
西施犬

Pekingese
北京狗

Cats of all sizes 大小不同的貓

Cats are also known as felines. Although they come in lots of different sizes, they share many of the same characteristics.

貓也被稱為貓科動物。儘管牠們體型大小各有不同，但全都具有很多共同特徵。

lynx
猞猁

colocolo
南美草原貓

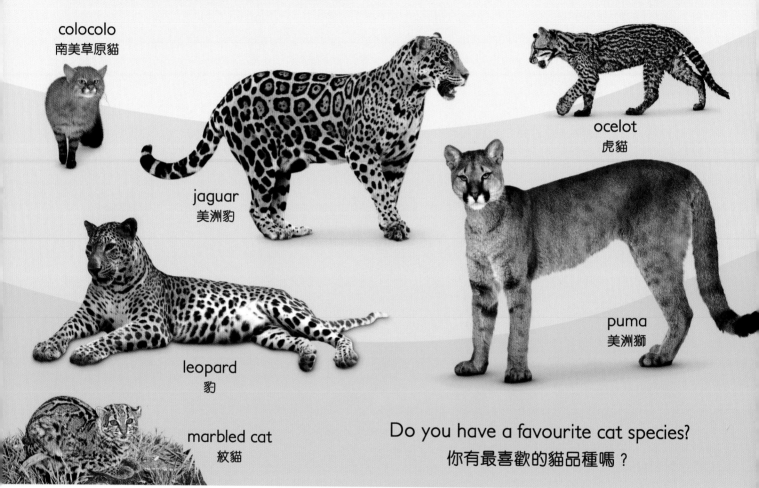

jaguar
美洲豹

ocelot
虎貓

leopard
豹

puma
美洲獅

marbled cat
紋貓

Do you have a favourite cat species?
你有最喜歡的貓品種嗎？

Feline features
貓科特徵

hunters
捕獵的本能

carnivorous
肉食性動物

curved claws
彎彎的爪

cheetah
獵豹

caracal
猁貓

sand cat
沙漠貓

serval
藪貓
(藪：粵音手)

Pallas's cat
兔猻
(猻：粵音宣)

snow leopard
雪豹

Domestic cat breeds
家貓品種

Himalayan
喜馬拉雅貓

Domestic
Shorthair
短毛家貓

Maine Coon
緬因貓

Siamese
暹羅貓

jaguarundi
細腰貓

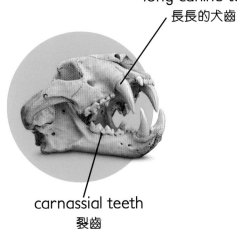

long canine teeth
長長的犬齒

carnassial teeth
裂齒

whiskers
鬍鬚

night vision
夜視能力

Animal sounds 動物聲音

Animals make noises for many reasons. Different sounds help them to communicate with family, scare away predators, and attract mates.

動物發出聲音的原因有很多。不同的聲音幫助牠們與同伴交流、嚇跑捕食者，或吸引配偶。

Habitats sound different because of the creatures that live there.

不同生境的環境聲音，會因為在那兒棲息的動物而不同。

Ocean 海洋

dolphins
海豚

click
嗒答聲

whistle
哨聲

orca
虎鯨

Countryside 田野

chirp
啁啾

nuthatch
五子雀

screech
嘎吱

barn owl
倉鴞
（鴞：響音集）

hoot hoot
嗚嗚

tawny owl
灰林鴞

bark
吠叫

fox
狐狸

scream
尖叫

peck peck
噠噠噠

squeak
吱吱

mouse
老鼠

croak
呱呱

woodpecker
啄木鳥

frog
青蛙

Savannah
稀樹草原

chatter
唧唧叫

vervet monkey
長尾猴

trumpet
叭噢

elephant
大象

rattle 嘎嘎

hiss
嘶嘶

snake
蛇

snarl
咧嘴吼叫

roar
怒吼

growl
咆哮

lion
獅子

Farmyard
農場

oink
嗯嗯

snort 哼哼

moo
哞哞

cow
乳牛

pig
豬

bray
驢叫

whinny
尖嘶

hee-haw
唏嗬

neigh
咻咻

donkey
驢

nicker
輕嘶

horse
馬

gobble gobble
咯咯叫

baaa
咩

bleat
羊叫聲

turkey
火雞

sheep
綿羊

quack
呱呱

hiss
嘶嘶

cockadoodledoo
咯喔咯咯

honk
嘎嘎

cluck
咯咯

goose
鵝

duck
鴨

chicken
雞

Garden
花園

woof
汪汪叫

miaow
喵喵

bark
吠叫

coo
咕咕

cat
貓

bee
蜜蜂

pigeon
鴿子

dog
狗

buzz
嗡嗡

Superpowers
動物的超能力

Some creatures have adaptations that are so incredible, they sound like superpowers!

有些生物的適應能力令人難以置信，聽起來就像有超能力！

firefly
螢火蟲

bioluminescent 生物發光

using
ultrasound
利用超聲波

little
brown bat
小棕蝠

run as fast
as a car
跑得跟汽車
一樣快

tardigrade
水熊蟲

hear sound
too high for
human ears
能聽高頻聲波

cheetah
獵豹

suspended animation
(pausing life)
假死（偽裝成死亡狀態）

plumed basilisk
雙嵴冠蜥

run on
water
在水上跑步

electric eel
電鰻

salmon
三文魚

give an electric shock
電擊

navigate long distances
長途跋涉

intelligent
聰明

flexible
伸縮自如

most
venomous sting
最毒的刺

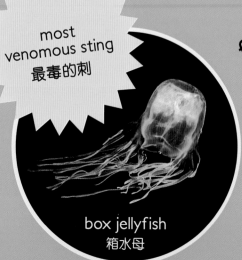

firefly squid
螢火魷

glow in the dark
在黑暗中發光

box jellyfish
箱水母

octopus
章魚

escape traps
逃出陷阱

world's
fastest animal
世上最快的動物

peregrine falcon
游隼

*quick
dive*
高速
潛水

gecko
壁虎

walk on ceilings
飛簷走壁

sticky toe pads
具吸附力的腳墊

amazing sense
of smell
驚人的嗅覺

most
venomous bite
最毒的一咬

climb up
walls
在牆上爬行

funnel web spider
漏斗網蜘蛛

jumping 跳躍

cat flea
貓蚤

Alpine ibex
羱羊
（羱：粵音原）

silvertip grizzly bear
灰熊

leap 150 times its
body length
跳躍身體長度的150倍

great balance
絕佳平衡力

punch
faster than
a bullet
出拳比子彈還快

peacock
mantis shrimp
雀尾螳螂蝦

On the move
移動方式

Animals move in different ways to get around, escape predators, or hunt prey.

動物以不同移動方式來四處走動、逃避捕食者或捕捉獵物。

Flying
飛行

soar
翔翔

flutter
翩翩起舞

flap
鼓翼

hover
盤旋

Hunting
捕獵

chase
追逐

stalk
跟蹤

prowl
徘徊

Slow moving 緩慢移動

crawl
爬行

belly crawl
腹部貼地爬行

walk
走路

In the water 在水中

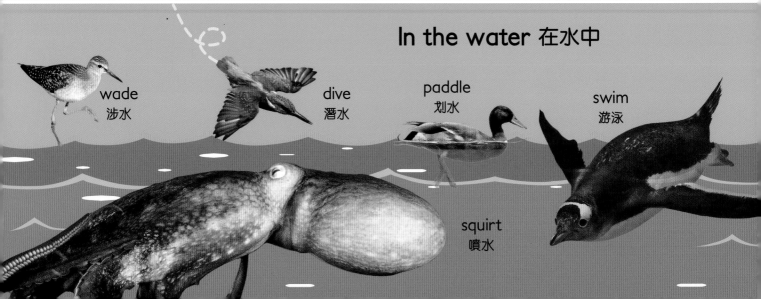

wade
涉水

dive
潛水

paddle
划水

swim
游泳

squirt
噴水

active flight
飛翔

glide
滑行

Quick movements
快速移動

jump
彈跳

hop
跳躍

gallop
奔馳

run
疾奔

loop
弓身走路

In the trees 樹木間

swing
盪來盪去

climb
爬

Unusual movements
奇特的移動

knuckle walk
指關節着地走路

roll
滾動

sidewind
側繞式移動

float
漂浮

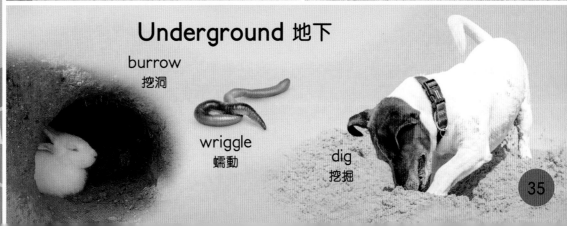

Underground 地下

burrow
挖洞

wriggle
蠕動

dig
挖掘

Unusual animals
獨特的動物

There are a wide variety of animals.
Some don't look like animals at all!

動物種類各式各樣，但有些看起來
實在太神奇了！

saiga antelope
高鼻羚羊

pink fairy armadillo
倭犰狳
（犰狳：粵音求餘）

northern three-toed
jerboa
三趾跳鼠

mata mata
sea turtle
瑪塔蛇頸龜

Satanic leaf-tailed gecko
角葉尾壁虎

star-nosed
mole
星鼻鼴

thorny devil
澳洲魔蜥

vase sponge
花瓶海綿

sea pig
海豬

warty
frogfish
大斑蟾魚
（蟾：粵音逼）

glass sponge
六放海綿

brain coral
腦珊瑚

purple sea pen
海鰓

mushroom coral
真葦珊瑚

limpet
帽貝

Honduran
white bat
洪都拉斯白蝙蝠

Indian
purple frog
印度紫蛙

greater lophorina
華美風鳥

magnificent
frigate bird
華麗軍艦鳥

hummingbird hawk-moth
蜂鳥鷹蛾

Japanese emperor
caterpillar
大紫蛺蝶幼蟲

dumbo
octopus
小飛象章魚

red-lipped
batfish
達氏蝙蝠魚

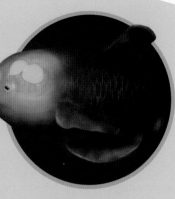

blue dragon sea slug
大西洋海神海蛞蝓

firework jellyfish
馬氏海生水母

Macropinna
microstoma
大鰭後肛魚

blobfish
水滴魚

goblin shark
歐氏尖吻鮫

barnacle
藤壺

leafy sea dragon
葉海龍

sea bunny
海兔

Venus fan
金星海扇

37

Colourful animals
色彩斑斕的動物

Vibrant colours often have a purpose in the animal kingdom, such as warnings, camouflage, and showing off!

鮮豔的顏色在動物世界中通常別有用途，例如警告、偽裝或炫耀！

I'm pink because of the food I eat!
我因為進食的東西而變成粉紅色。

flamingo
紅鸛

keel-billed toucan
彩虹巨嘴鳥

neon tetra
霓虹燈魚

freshwater angelfish
淡水神仙魚

scarlet macaw
緋紅金剛鸚鵡

My bright colours help me attract a mate.
我亮麗的顏色幫助吸引配偶。

crowntail betta
獅王鬥魚

fiery throated hummingbird
火喉蜂鳥

golden snub-nosed monkey
金絲猴

Moorish idol
鐮魚

Indian peafowl
藍孔雀

candy basslet
卡氏長鱸

oriental dwarf kingfisher
三趾翠鳥

Wilson's bird-of-paradise
威氏麗色風鳥

green discus
塔爾氏盤麗魚

boomslang snake
非洲樹蛇

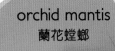

orchid mantis
蘭花螳螂

My colour helps me hide among flowers.
我的顏色幫助我藏身在花叢中。

blue morpho
大藍閃蝶

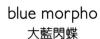

Indonesian pit viper
島竹葉青

peacock spider
孔雀蜘蛛

eastern box turtle
東部箱龜

panther chameleon
豹變色龍

red lionfish
魔鬼簑鮋
（簑：粵音梳）

peacock mantis shrimp
雀尾螳螂蝦

My colour warns other animals that I'm poisonous.
我的顏色警告其他動物我有毒！

crowned jellyfish
皇冠水母

blue dart frog
藍箭毒蛙

jewel beetle
寶石甲蟲

Picasso bug
畢卡索盾蝽

rose maple moth
玫瑰楓葉蛾

mandrill
山魈
（彩面狒狒）

golden tortoise beetle
黃金金龜子

rainbow parrotfish
虹彩鸚嘴魚

blue-ringed octopus
藍環章魚

beadlet anemone
等指海葵

royal gramma
紫天堂

39

Camouflage
隱身術

Camouflage helps animals to blend in with their habitats and stay hidden from view.

偽裝有助動物融入生境，好好隱藏。

white-tailed ptarmigan
白尾雷鳥

look like snow
偽裝成白雪

Arctic fox
北極狐

Arctic 北極

look like bark and leaves
偽裝成樹皮和樹葉

peppered moth
樺尺蛾

copperhead snake
銅頭蝮

great horned owl
大鵰鴞

long-eared owl
長耳鴞

Boreal forest 寒溫帶針葉林

look like plants
偽裝成植物

mossy leaf-tail gecko
苔蘚葉尾壁虎

ghost mantis
幽靈螳螂

walking stick
竹節蟲

common potoo
普通林鴟
（鴟：粵音雌）

(40) ## Tropical forest 熱帶雨林

look like trees and tree bark
偽裝成樹木和樹皮

red squirrel
紅松鼠

Eastern chipmunk
東美花鼠

mountain caribou
高山馴鹿

Temperate forest 溫帶森林

Savannah 稀樹草原

look like dappled shade
偽裝成斑駁的陰影

spotted hyena
斑點鬣狗
（鬣：粵音獵）

impala
高角羚

leopard
豹

African wild dog
非洲野犬

Underwater 水中

blue shark
大青鯊

hard to see from above and below
難以從水面或水底辨認

look like seafloor
偽裝成海牀

stone flounder
石鰈

Flowers 花叢

look like flowers
偽裝成花朵

golden rod crab spider
秋麒麟蟹蛛

orchid mantis
蘭花螳螂

Riverbank 河岸

nightjar
夜鷹

look like the ground
偽裝成地面

mugger crocodile
恆河鱷

look like a log
偽裝成木頭

Home sweet home 温馨家園

There are all kinds of homes in the animal kingdom. Some creatures build their own home, while others find existing spots to settle in.

動物世界有各式家園。一些生物會建造自己的家園，
而另一些則尋找現成的藏身之所安頓下來。

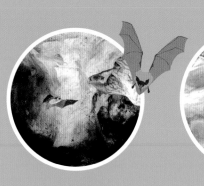

bat 蝙蝠
roost 蝙蝠洞

bear 熊
den 洞穴

wild boar 野豬
**temporary nest
臨時巢穴**

prairie dog 土撥鼠
town 如地下城的洞穴

harvest mouse 歐洲田鼠
nest 巢窩

red squirrel 紅松鼠
drey 松鼠窩

otter 水獺
holt 林地巢穴

beaver 河狸
**lodge inside a dam
建水壩為巢穴**

fox 狐狸
earth 地洞

rabbit 兔子
warren 兔子洞窟

water vole 水䶄
burrow 水邊洞穴

badger 獾
sett 獾洞

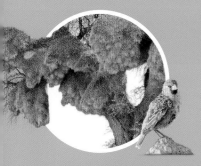

sociable weaverbird
羣織雀
multistorey nest 複合巢

baya weaverbird
黃胸織布鳥
nest colony 葫蘆形鳥巢

rufous hornero 棕灶鳥
clay nest 麵包爐形鳥巢

golden eagle 金雕
eyrie 鷹巢

clownfish 小丑魚
anemone 海葵

limpet 帽貝
rock 岩石

hanging
nest
吊巢

Montezuma oropendola 褐色擬椋鳥
pendulous nest 懸垂的鳥巢

nest with
hexagonal
cells
六角形蜂房

paper wasp
造紙胡蜂
nest 蜂巢

curl a
leaf to
make a
home
捲葉為家

Australian leaf-curling spider
澳洲捲葉蛛
leaf 葉子

froghopper 沫蟬
cuckoo spit 泡沫

European red wood ant
歐洲紅木蟻
anthill 蟻冢

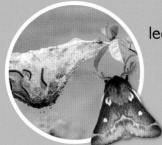

tent moth 天幕毛蟲蛾
tent made of silk
幼蟲吐絲成巨網

leaves woven
together
樹葉交織
在一起

Australian weaver ant
澳洲編織蟻
nest made of leaves 以樹葉築巢

Baby animals 動物寶寶

Animals change as they grow up. Some change completely, others just grow bigger.

動物長大後會發生變化，有些外觀完全改變，有些只是體型變大。

hippopotamu
河馬

whale
鯨

chicken
小雞

Chicks
雛鳥

penguin
企鵝

Calves
大型哺乳類
幼崽

kid
小山羊

foal
幼馬

fry
魚苗

cheetah
獵豹

hatchling
孵化

Baby
animals
動物寶寶

wildebeest
牛羚

piglet
小豬

lamb
小綿羊

fawn
小鹿

Cubs
幼獸

wombat
袋熊

Joeys
有袋類
幼崽

kangaroo
袋鼠

opossum
負鼠

panda
熊貓

Green sea turtle
綠蠵龜（蠵：粵音攜）

**reptiles, fish, birds :
egg laying**

爬行動物、魚、雀鳥：
下蛋

ɑying eggs
下蛋

pit in sand
沙坑

Otter
水獺

**mammals :
live young**

哺乳動物：幼崽

pup
水獺寶寶

live birth
胎生

Frog
青蛙

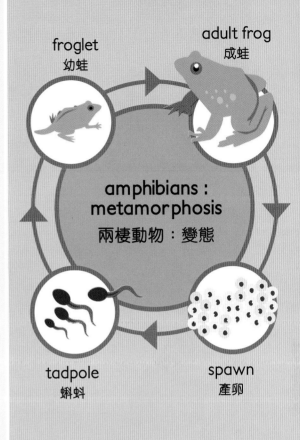

froglet
幼蛙

adult frog
成蛙

**amphibians :
metamorphosis**

兩棲動物：變態

tadpole
蝌蚪

spawn
產卵

Spurge hawk-moth
大戟鷹蛾

pupa 蛹

adult 成蛾

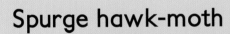

**insects :
complete
metamorphosis**

昆蟲：完全變態

larva
幼蟲

egg
蛾卵

Green shield bug
紅尾碧蝽

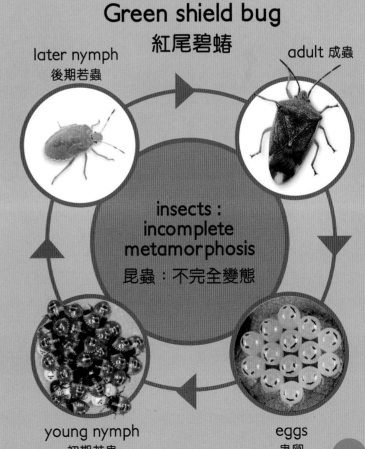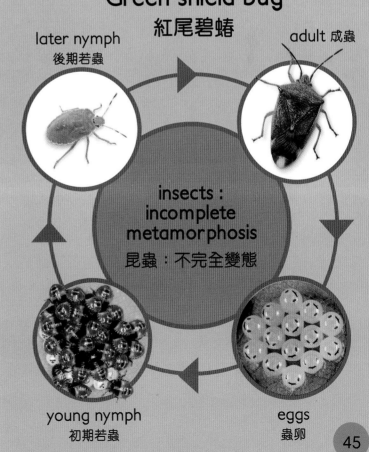

later nymph
後期若蟲

adult 成蟲

**insects :
incomplete
metamorphosis**

昆蟲：不完全變態

young nymph
初期若蟲

eggs
蟲卵

Eggs of all kinds
各種各樣的蛋

Fish eggs, bird eggs, turtle eggs, insect eggs – they all look different.

魚卵、鳥蛋、龜蛋、昆蟲卵，
它們看起來都不一樣。

Birds
鳥類

chicken
雞蛋

emu
鴯鶓蛋

vervain
hummingbird
小吸蜜蜂鳥蛋

ostrich
駝鳥蛋

melodious
warbler
歌鸲鶯蛋

Reptiles
爬行動物

lizard
蜥蜴蛋

sea turtle
海龜蛋

crocodile
鱷魚蛋

Amphibians
兩棲動物

newtspawn
水螈卵

toadspawn
蟾蜍卵

surface of the water
在水面

frogspawn
青蛙卵

Octopuses and squids
章魚和魷魚

opalescent market squid
乳光槍烏賊卵

veined octopus
條紋蛸卵

Fish 魚類

horn shark
角鯊卵

salmon 三文魚卵

lesser spotted dogfish
小點貓鯊卵

sturgeon
鱘魚卵

Molluscs
軟體動物

snail
蝸牛卵

Insects 昆蟲

monarch butterfly
帝王斑蝶卵

spined soldier bug
斑腹刺益蝽卵

male carries eggs on its back
雄性背着卵

giant water bug
巨型水蟲卵

ladybird
瓢蟲卵

owl butterfly
貓頭鷹環蝶卵

Myriapods
多足類動物

red-headed centipede
赤蜈蚣

Arachnids
蛛形綱動物

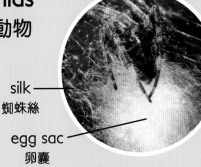

silk
蜘蛛絲

European garden spider
十字園蛛

egg sac
卵囊

Incredible bodies
神奇的身體

Look at all the parts which make up these amazing animals.
看看這些奇妙動物的身體各個部分。

Skeletons
骨骼

skeleton muscles
骨骼   肌肉

exoskeleton
（muscles inside
the skeleton）
外骨骼型(骨骼包裹着肌肉)

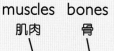

muscles bones
肌肉   骨

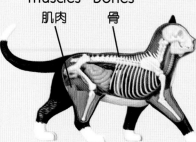

endoskeleton
（muscles outside
the skeleton）
內骨骼
（肌肉包裹着骨骼）

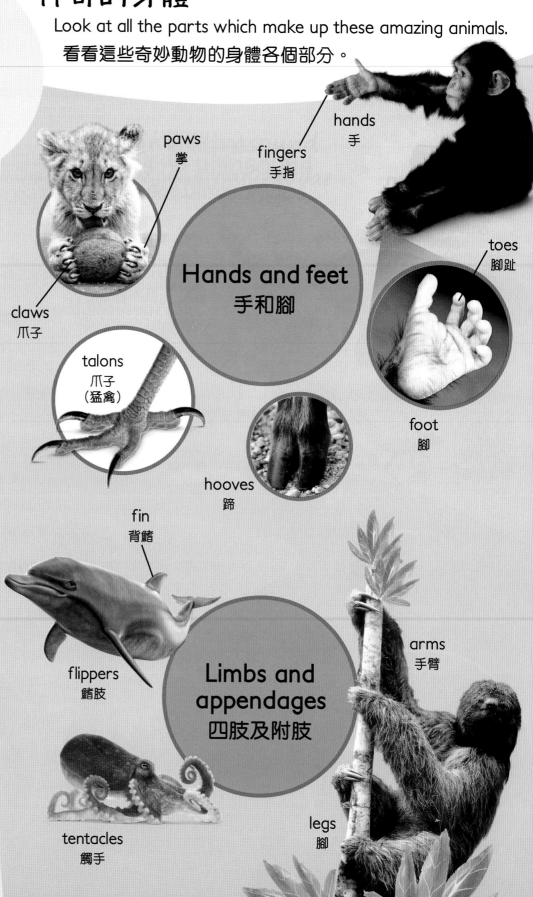

paws
掌

fingers
手指

hands
手

claws
爪子

Hands and feet
手和腳

toes
腳趾

talons
爪子
（猛禽）

foot
腳

hooves
蹄

fin
背鰭

arms
手臂

flippers
鰭肢

Limbs and
appendages
四肢及附肢

tentacles
觸手

legs
腳

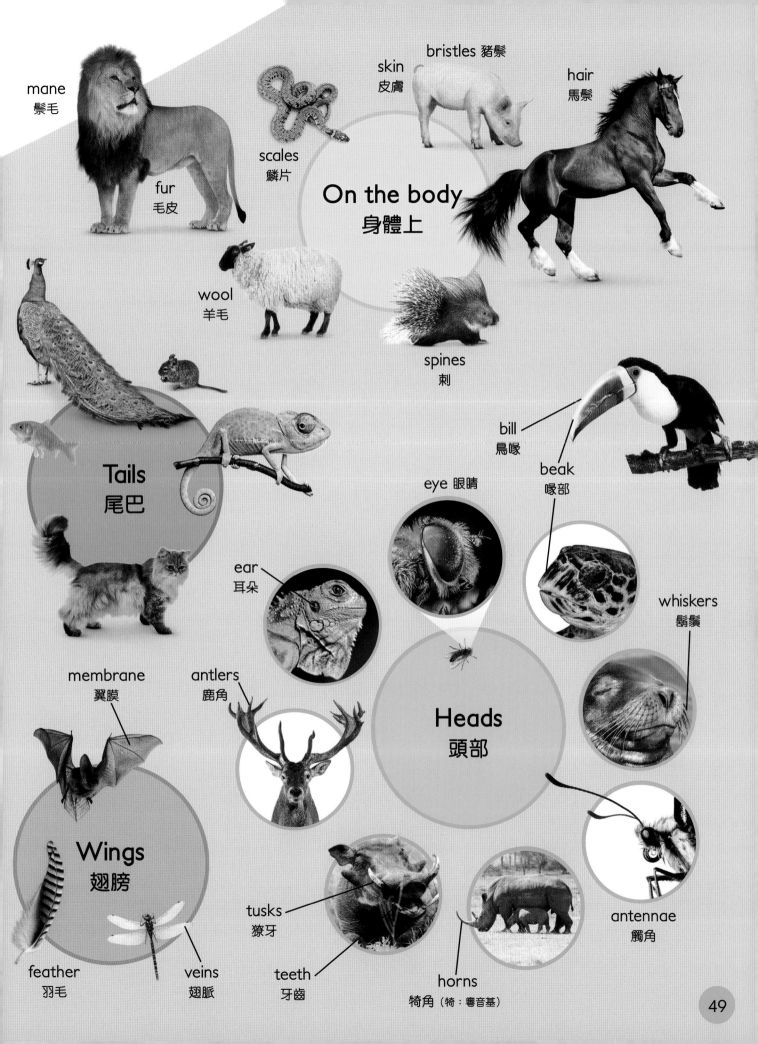

mane
鬃毛

scales
鱗片

skin
皮膚

bristles 豬鬃

hair
馬鬃

fur
毛皮

**On the body
身體上**

wool
羊毛

spines
刺

bill
鳥喙

beak
喙部

**Tails
尾巴**

eye 眼睛

ear
耳朵

whiskers
鬍鬚

membrane
翼膜

antlers
鹿角

**Heads
頭部**

**Wings
翅膀**

antennae
觸角

tusks
獠牙

feather
羽毛

veins
翅脈

teeth
牙齒

horns
犄角（犄：粵音基）

49

Defending 自衞技能

Animals have lots of different ways to protect themselves.
動物有很多不同的方法來保護自己。

hornet
大黃蜂

crown of
thorns starfish
棘冠海星

porcupine
箭豬

millipede
千足蟲

ball python
球蟒

hedgehog
刺猬

Spikes
棘刺

Curling up
蜷曲

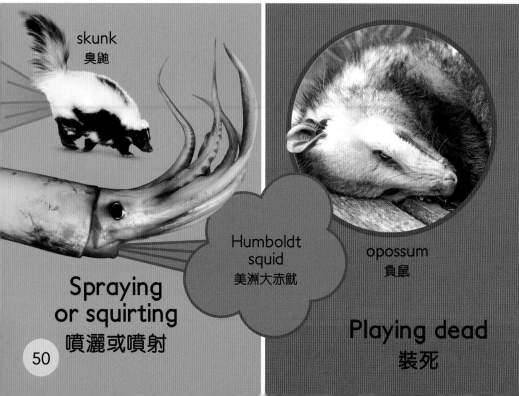

skunk
臭鼬

Humboldt
squid
美洲大赤魷

opossum
負鼠

Spraying
or squirting
噴灑或噴射

Playing dead
裝死

crested gecko
睫角壁虎

stone
crab
石蟹

Dropping
a limb or tail
掉下肢體或尾巴

50

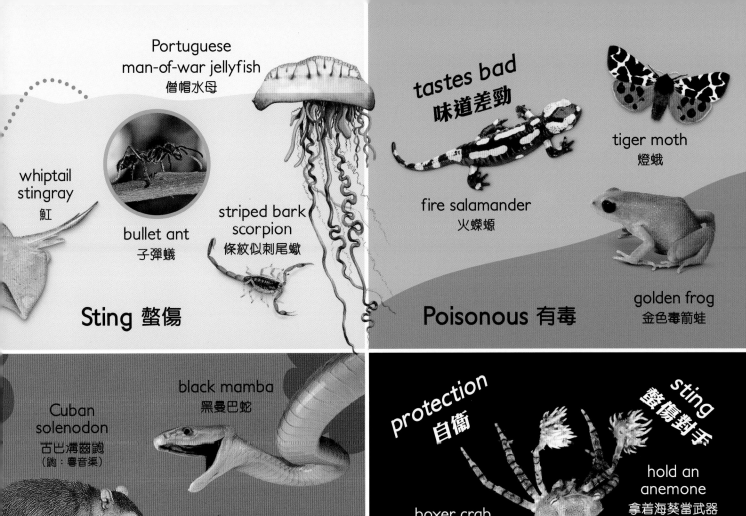

Portuguese
man-of-war jellyfish
僧帽水母

whiptail
stingray
魟

bullet ant
子彈蟻

striped bark
scorpion
條紋似刺尾蠍

Sting 螫傷

tastes bad
味道差勁

tiger moth
燈蛾

fire salamander
火蠑螈

Poisonous 有毒

golden frog
金色毒箭蛙

black mamba
黑曼巴蛇

Cuban
solenodon
古巴溝齒鼩
（鼩：粵音渠）

dangerous
危險

Venomous bite 毒咬

protection
自衛

sting
螫傷對手

boxer crab
拳擊蟹

hold an
anemone
拿着海葵當武器

**Using another animal
利用其他生物**

armadillo 犰狳

starlings
椋鳥

schools of fish
魚羣

tortoise beetle
龜金花蟲

Indian
pangolin
印度穿山甲

meerkats
狐獴

**Gathering in groups
聯羣結隊**

**Armour plates
覆蓋硬殼**

51

Carnivores and herbivores

肉食性動物和草食性動物

Carnivores eat other animals. Herbivores only eat plants.
Omnivores have adapted to eat both meat and plants.

肉食性動物會吃其他動物；草食性動物只吃植物；雜食性
動物則吃肉類和植物。

binocular vision
雙眼視覺

buzzard skull
禿鷹顱骨

talons
鷹爪

tongue
舌頭

incisors
門齒

sharp
鋒利

claws
爪子

wolf skull
狼顱骨

teeth
牙齒

canine
犬齒

rows of teeth
上下排牙齒

ladybird
瓢蟲

slice
撕開

**Carnivores
肉食性動物**

baleen
鯨鬚

Can you name any other
carnivorous animals?

你能說出其他肉食性
動物的名字嗎？

filter feeder
濾食性動物

distensible jaw
可擴張的顎部

bear
熊

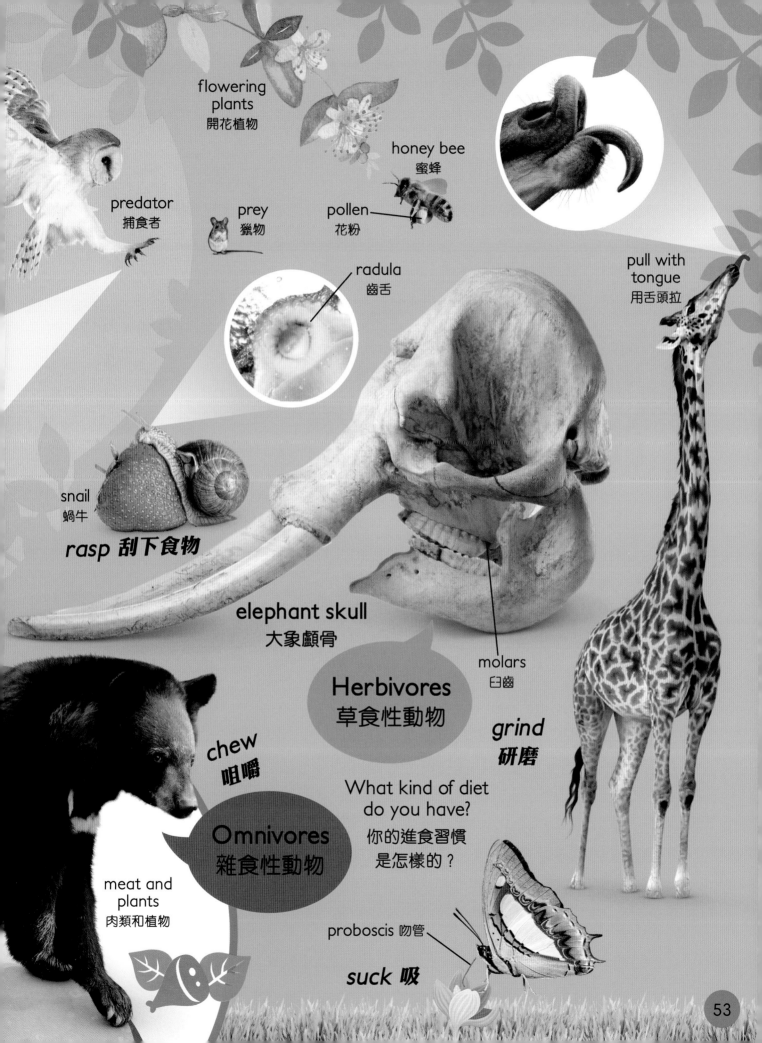

flowering plants
開花植物

honey bee
蜜蜂

predator
捕食者

prey
獵物

pollen
花粉

radula
齒舌

pull with tongue
用舌頭拉

snail
蝸牛

rasp 刮下食物

elephant skull
大象顱骨

molars
臼齒

Herbivores
草食性動物

grind
研磨

chew
咀嚼

What kind of diet do you have?
你的進食習慣是怎樣的？

Omnivores
雜食性動物

meat and plants
肉類和植物

proboscis 吻管

suck 吸

Animal diaries

動物日記

Discover what animals get up to at different times of the day and night.

一起來了解動物在晝夜的不同時間都在做什麼。

Tiger 老虎

sleep in the shade
在陰涼處睡覺

snooze
打瞌睡

Iguana 鬣蜥

diurnal (active in the day)
晝行性（活躍於白天）

bask to warm up
曬太陽暖身

feed on leaves and fruit
進食葉和水果

protect territory
保護領地

Barn owl 倉鴞

dawn
黎明

hunt
狩獵

sleep
睡覺

crepuscular (active at dawn and dusk)
晨昏活動型
（活躍於黎明和黃昏）

Morning 早上

Afternoon 下午

nocturnal
(active at night)
夜行性
（活躍於夜間）

patrol territory
巡邏領地

hunt
狩獵

scent-marking
標記氣味

escape from predators
逃離捕食者的狩獵

find a spot to rest
找個地方休息

inactive because it is colder
天氣冷要休息

feed
覓食

dusk
黃昏

roost in trees or empty buildings
棲息在樹上或閒置的建築物中

Evening 傍晚

Night 晚上

People and animals
人類與動物

There are many careers that involve working with animals.

有許多職業都需要與動物打交道。

zookeeper
動物園管理員

wildlife ranger
野生動物護林員

zoologist
動物學家

zoo
動物園

horse trainer
練馬師

research
從事研究

marine biologist
海洋生物學家

camera person
攝像師

wildlife presenter
野生動物導賞員

photographer
攝影師

study
研究

naturalist
博物學家

bacteria
細菌

entomologist
昆蟲學家

camera
相機

Would you like to work with animals when you grow up?
你長大後想和動物一起工作嗎？

operate
設備操作

vaccinate
注射疫苗

vet
獸醫

jockey
騎師

groom
馬夫

rescue
拯救行動

rehabilitate
康復治療

veterinary nurse
獸醫護士

animal welfare officer
動物福利員

dog groomer
犬隻美容師

virus
病毒

dog trainer
馴犬師

virologist
病毒學家

beekeeper
養蜂人

feed
餵飼

microbiologist
微生物學家

honeycomb
巢脾
（脾：粵音皮）

hive
蜂箱

farmer
農夫

57

Endangered or at risk
瀕危或處於危險

These animals are badly affected by human activities. Unless people try to help, they could become extinct.

這些動物受到人類活動的嚴重影響，除非人們盡力提供幫助，否則牠們會趨向滅絕。

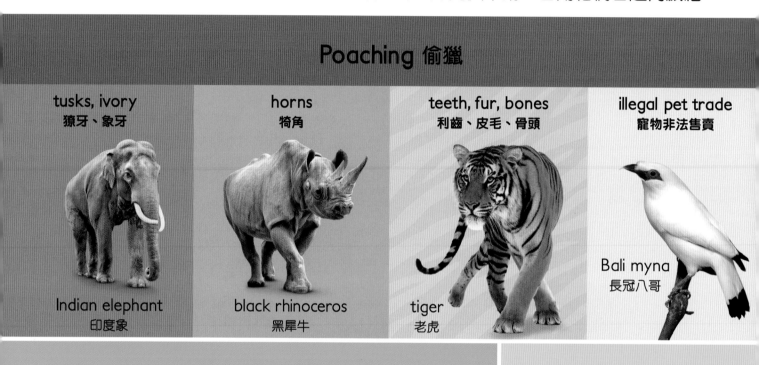

Poaching 偷獵

tusks, ivory
獠牙、象牙

Indian elephant
印度象

horns
犄角

black rhinoceros
黑犀牛

teeth, fur, bones
利齒、皮毛、骨頭

tiger
老虎

illegal pet trade
寵物非法售賣

Bali myna
長冠八哥

Climate change and global warming
氣候轉變及全球暖化

sea temperature rising
海水溫度上升

elkhorn coral
鹿角珊瑚

bush fires
叢林大火

koala
樹熊

melting ice
冰層融化

polar bear
北極熊

Invasive species
物種入侵

population decline
數量遞減

disease brought by grey squirrels
由灰松鼠帶來的疾病

European red squirrel
歐亞紅松鼠

Disease 疾病

African wild dog
非洲野犬

Overhunting and overfishing
過度捕獵與捕撈

loss of food
失去食物

Adélie penguin
阿德利企鵝

Atlantic bluefin tuna
大西洋藍鰭吞拿魚

green sea turtle
綠蠵龜

Habitat loss 喪失生境

more buildings for humans
越多人類建築

less habitat for wildlife
越少野生動物生境

urban sprawl
城市蔓延

Lange's metalmark butterfly
花蜆蝶
（蜆：粵音顯）

Florida panther
佛羅里達美洲獅

loggerhead sea turtle
赤蠵龜

deforestation
濫伐林木

cutting down trees
砍樹

orangutan
紅毛猩猩

Canadian caribou
加拿大馴鹿

mountain gorilla
山地大猩猩

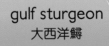

gulf sturgeon
大西洋鱘

fewer bodies of water
水體減少

Insecticides 殺蟲劑

chemicals can kill animals or make them sick
化學品會殺死動物或使牠們生病。

San Joaquin kit fox
敏狐

Crotch's bumblebee
克羅奇大黃蜂

Extinct species
滅絕物種

Some animals are unable to adapt as their environment changes, and they become extinct. This means the species has no living members.

有些動物無法適應環境的變化，走向滅絕。
這意味該物種再沒有存活的成員。

sea scorpions
海蠍

Spinosaurus
棘龍

giant wombat
雙門齒獸

sabre-toothed tiger
劍齒虎

Humans often cause the environmental changes that lead to animal extinction.

人類經常引起導致動物滅絕的環境變化。

dodo
渡渡鳥

auroch
原牛

Tasmanian tiger
袋狼

crescent nail-tail wallaby
新月甲尾袋鼠

Rocky Mountain locust
洛磯山黑蝗

passenger pigeon
旅鴿

Xerces blue butterfly
加利福尼亞甜灰蝶

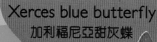

pig-footed bandicoot
豚足袋狸

Polynesian tree snail
玻里尼西亞樹蝸牛

New Zealand grayling
尖吻南茴魚

Brachiosaurus
腕龍

Pteranodon
無齒翼龍

Iguanodon
禽龍

Ankylosaurus
甲龍

Allosaurus
異特龍

Dinosaurs may have become extinct when a meteorite collided with the Earth.
恐龍可能因為隕石撞擊地球而滅絕。

woolly mammoth
猛獁象

great auk
大海雀

eastern elk
加拿大馬鹿

Steller's sea cow
大海牛

Caribbean monk seal
加勒比僧海豹

Pyrenean ibex
庇里牛斯羱羊

West African black rhinoceros
西部黑犀

Mythical creatures
神話生物

Some animals exist only in stories. Every culture has its own stories and its own mythical beasts.

有些動物只存在於神話故事中。每種文化都有自己的神話和神獸。

faun
牧神法翁

fairy
仙女

Hydra
海德拉獸

Pegasus
飛馬

werewolf
狼人

Sasquatch (Bigfoot)
大腳怪

Kun-Peng
鯤鵬

centaur
半人馬

dragon
龍

leprechaun
矮妖

Cerberus
刻耳柏洛斯 (地獄三頭犬)

Jiuwei Hu
九尾狐

Zouwu
騶吾

long
Chinese dragon
中國龍

mermaid
美人魚

manticore
蠍獅

Which of these mythical
creatures do you recognise?
你認識哪些神話生物？

unicorn
獨角獸

sphinx
斯芬克斯（獅身人面獸）

troll
山精

kraken
挪威海怪

Ao 鰲

elf 小精靈

Pixiu 貔貅
（粵音皮休）

minotaur
米洛陶
（牛頭人）

Chi 螭
（粵音雌）

griffin
獅鷲

phoenix
鳳凰

harpy
哈比
（鷹身女妖）

Shang-Yang
喬羊

yeti
雪人

Loch Ness Monster
尼斯湖水怪

Acknowledgements 鳴謝

謹向以下人員致謝，感謝他們在編輯本書上提供的協助：
Sif Nørskov (編輯)、Sophie Parkes (編輯) 以及 Polly Goodman (校對)

謹向以下單位致謝，感謝他們允許使用照片：
(Key: a-above; b-below/bottom; c-centre; f-far; l-left; r-right; t-top)

123RF.com: Andrzej Tokarski / ajt 64br, Anan Kaewkhammul / anankkml 28ca, Benjamin King / benjaminjk 51ca, bonzami emmanuelle 20c, Corey A Ford 61tc, Duncan Noakes / fouroaks 17ca, Eric Isselee 7cb, 50cl, 50clb, Eric Isselee / isselee 1bl, 28cr, 29cra, 58cb, Anan Kaewkhammul 28cl, max5128 16cra, Ben McRae 41cla (texture), nrey 2tr, Andrei Samkov / satirus 52c, smileus 44bl, swavo 6cra (glass), 11cl (glass), 32cla (glass), Thawat Tanhai 38clb (Kingfisher), Nicholas Toh 37bl, Pavlo Vakhrushev / vapi 9cb, 64crb, Oleg Znamenskiy zov666@gmail.com 41ca; **Alamy Stock Photo:** AGAMI Photo Agency / Andy & Gill Swash 40bc, Linda Freshwaters Arndt 34cla, Art Collection 3 60crb (wallaby), Avalon.red / Anthony Bannister 11cr, Avalon.red / Stephen Dalton 35tc, Biosphoto / Adam Fletcher 39cla, Biosphoto / Sergio Hanquet 51clb, Biosphoto / Sylvain Cordier 42ca (nest), blickwinkel / AGAMI / J. Eaton 36cla, blickwinkel / F. Hecker 42c, 45bc, blickwinkel / F. Teigler 11bc, blickwinkel / H. Bellmann / F. Hecker 45br, blickwinkel / Lundqvist 40clb, Buiten-Beeld / Jelger Herder 14clb, Nigel Cattlin 11ca, cbstockfoto 43cb, Clarence Holmes Wildlife 47clb, Corbin17 38bl, Rick Dalton - Wildlife 59ca, Design Pics Inc / Alaska Stock RM / Thomas Kline 47cl, Digital Arts Pro 48cr, Reinhard Dirscherl 37c, 39bc, David Fleetham 20crb (puffer), 51cr, Florilegius 60crb (bandicoot), FLPA 26cb (Elkhound), 47tr, 61crb, Bill Gozansky 40bl, Frank Hecker 21clb, Louise Heusinkveld 54bc (sleep), Imagebroker / Arco / G. Lacz 29crb, imageBROKER / Dirk Funhoff 15ca, imageBROKER / Gerry Pearce 17c, 33c, imageBROKER / Michaela Walch 43tr, imageBROKER / R. Dirscherl 36bl, Juniors Bildarchiv GmbH / Arndt, S.E. / juniors@wildlife 55bc, Ivan Kuzmin 10bl, mike lane 19br, M@rcel 43br, Francisco Martinez-Clavel Martinez 10c, Chris Mattison 41cb, mauritius images GmbH / BY 25c, mauritius images GmbH / Solvin Zankl 46bc, McPhoto / Rolf Mueller 17tr, Mic Clark Photography 55cra, Minden Pictures / Norbert Wu 20bl, 47ca (horn), Natural History Museum, London 60clb, 61cb, Nature Photographers Ltd / Paul R. Sterry 8clb (Shrimp), 20br, 32clb, 43bc (moth), 47tc, Nature Picture Library 31cl, 50bl, Nature Picture Library / 2020VISION / Alex Mustard 21c, Nature Picture Library / Bence Mate 32crb, Nature Picture Library / Chris Mattison 15cra, Nature Picture Library / Eric Medard 42cla, Nature Picture Library / MYN / Joris van Alphen 15c, Nature Picture Library / MYN / Lily Kumpe 10cr, Nature Picture Library / MYN / Marc Pihet 43fclb, Nature Picture Library / Nick Upton 42bc, Andrey Nekrasov 16br, NOAA 37cl, Matteo Omied 59tr, Panoramic Images 49bc, Papilio / Robert Pickett 47crb (Owl), PhotoStock-Israel / Alon Meir 44cl, Picture Partners 47ca, Adisha Pramod 37cr, 37clb, Gillian Pullinger 42bc (warren), Lee Rentz 14cl, Remo Savisaar 42bl, SBS Eclectic Images 18bl, Robert Scholl 22bl, steeve-x-art 60bl, Marko Steffensen 37crb, Stefan Sutka 28bl, Tom K Photo 58cl, Dave Watts 18ca, WhiskeyWolf 61br, WILDLIFE GmbH 21crb, Ray Wilson 12ca (albatross); **Ardea:** Danita Delimont / Kevin Schafer 48br, M. Watson 45cla; **J. Buys:** 47bl; **Depositphotos Inc:** DedMorozz 54ca, Nataly-Nete 35ca (Grass), sophyphotos 6bc; **Dorling Kindersley:** Jerry Young 8clb, 64tl, Gary Ombler / Cotswold Wildlife Park 17cla, Neil Fletcher 43fcrb, Terry Goss 21bl, Brian Gratwicke 32crb (eel), Jon Hughes 61cla, Barnabas Kindersley 49fbl, Liberty's Owl, Raptor and Reptile Centre, Hampshire, UK 40ca, Richard Ling 9tl, Prof. Marcio Motta 28cla, Colin Keates / Natural History Museum, London 40tc, 46br, 49bl, Frank Greenaway / Natural History Museum, London 8c, 51tr, 54bc, Gary Ombler / Natural History Museum 29bl, Karl Shone / Natural History Museum, London 23cl, Peter Chadwick / Natural History Museum, London 52cra, Linda Pitkin 41tr, Gary Ombler / Royal Botanic Gardens, Kew 64br (leaves), Harry Taylor The Natural History Museum, London 6bl, Dave King / Whipsnade Zoo, Bedfordshire 17tr (bear), Wildlife Heritage Foundation, Kent, UK 29clb, Jerry Young 2bl, 10clb, 17bc, 38ca, 39br; **Dreamstime.com:** 3drenderings 8clb (woodlouse), Adchariya 43fbl, Anastasiya Aheyeva 42cb (otter), Alfotokunst 17crb, Alle 57cb, 57bc, Alptraum 15bl, Alslutsky 56bc, Carlos Alvarez 27clb, Alxhar 48cla, Amwu 1cl, 25tl, 35crb, Anders93 47crb, John Anderson 58clb, Andylid 24clb, Amy Harris / Anharris 53r, Anitasstudio 53cl, Rafael Ben Ari 36crb, Andrey Armyagov 7br, Atalvi 31cla, Bouke Atema 13crb, Auris 36-37, Natalia Bachkova 35clb, Jason W. Baker 10crb, Belizar 51cb (Amradillo), Christopher Bellette 6cra (spider), Ben 38cb, John Biglin 42br, Lukas Blazek 29ca, Blueringmedia 35c (branch), 63cla, Anna Bocharova 25tc, Linda Bucklin 1br, 16b, 60crb, Mariusz Bugno 12cla, Neil Burton 35tc (hare), Steve Byland 7tc, 12br, 13bc, Martin Capek / Cappan 32crb (lighting), Carolinemaryan 24cra, Vladimir Cech 18cla, Chernetskaya 25br (tray), Chuotnhatdesigner 34-35b, Conchasdiver 36fbl, Rudmer Zwerver / Creativenature1 14bc, Brett Critchley 48cb, 49c, Cynoclub 11cra, Damedeeso 35br, Olga Demchishina 33bl, Nikolay Denisov 36tr, Dikkyoesin 39tr, Dave Massey / Dmass 53ca, Dndavis 39ca, Dennis Donohue 12cra, Dragoneye 13cla, 38fcra, Dwiputra18 39cb, Ian Dyball 33tl, Ecophoto 19cr, Dirk Ercken 15clb, 25cl, Evcrow 35b, Farinoza 24crb, 36clb, 39cb (frog), 49ca (spines), Melinda Fawver 39clb (Moth), Feathercollector 41cra, Ricardo De Paula Ferreira 43tc (Rufous), Iakov Filimonov 26cra, 40br, Fireflamenco 2fbl, 63cr, Svetlana Foote 32cr, 58c, Corey A Ford 60c, FotoJagodka 26clb, Martin Fredskov 42cb, Robert Fullerton 40cla, Gallinagomedia 12crb, Svetlana Gladkova 54cl, Godruma 42fbr, Steve Gould 6crb, Igor Groshev 42cr, Pascal Halder 33r, David Havel 39bc (anemone), 41br, Hellmann1 56l (Bark), Nynke Van Holten 29cr, Brett Hondow 35cl, Hsagencia 35bc (worm), Boonchuay Iamsumang 34cr, Icefront 13clb, Idreamphotos 45tl, Imagine98 49crb (whiskers), Inarik 34c, Irisbraunphotography 44crb, Eric Isselee 27ca (collie), Isselee 1ca, 8cr, 11cb, 12cr, 13ca, 13cl, 13c, 13cb (tit), 13bl, 13br, 14br, 17crb (koala), 18ca (tamarin), 19tc, 20cra, 20crb (Clownfish), 22c, 23tr, 23b, 24bl, 24bc, 25tc (gecko), 25cr, 26bl, 27tc, 27tr, 27cra, 27cb, 27bc, 29tc, 29cla, 30cb, 31tr, 31cr, 31clb, 34cl, 34bc, 35tl, 39cra, 39clb, 40tr, 40cb, 42cra (boar), 42clb, 42bl (Rabbit), 42fbl, 44cb (fawn), 45tc, 45ca, 45clb (x2), 45cb, 45bl, 49ca, 49cb (deer), 49cb (fly), 54-55ca, 55clb, 56cl, 58br, Maria Itina 19cl, Iakov Filimonov / Jackf 59c, JaCrispy 24cla, Jacsdreamjam 46bl, Jagodka 27br, James Group Studios, Inc. 25bl (cage), Jeff Jarrett 43clb, Jessamine 1crb, Jezbennett 19tc (Lemur), Jianghongyan 9tc, Jocrebbin 52b, Johannes Gerhardus Swanepoel / Johan63 41tc, Johannesk 35crb (Sand), 38crb, Angela Jones 35cb, Josefpittner 40cla (fox), Juliengrondin 54br, 55cb (Oak), Acharaporn Kamornboonyarush 10bc, Karelgallas 52-53bc, John Kasawa 34-35cb, Kateleigh 30crb, Elena Kazanskaya 25bl (bell), Alexia Khruscheva 49tr, Khunaspix 31crb, Liliia Khuzhakhmetova 16ca, Miroslaw Kijewski 40clb (mantis), Aleksei Kondraiuk 34bl, Natalia Korotaeva 32br, Vasily Kovalev 25fcr, Irina Kozhemyakina 10cb (x2), 17cl, Anna Kravchuk 34clb (Sandpiper), Tomas Krist 34cra, Matthijs Kuijpers 24tr, 39tl, 51c, Olga Kurbatova 57b, Alexey Kuznetsov 26ca, Erik Lam 27ca, 27bl, Lebedinski 26cb, Peter Lindholm 6c (Crayfish), Liumangtiger 13clb (macaw), Luayana 40br (texture), Thomas Lukassek 43c (limpet), Lunamarina 18cl, 20cb, Tono Balaguer / Lunamarina 59clb, Anton Lunkov 50cla, Yurii Lysiak 23cr, Macrovector 62tr, 62ca, 62c, 62cr, 62bl, 63 (x6), 63cb, 63cb (Minotaur), Cosmin Manci 10cl, Marcouliana 10ca, Marish 63clb, Marquise132 47cr, Martinlisner 20crb, Sutisa Kangvansap / Mathisa 53cb, Vaclav Matous 16crb, Aliaksandr Mazurkevich 19bl, Mikelane45 12c, 13cr, 18cb, 42cl, 42br (vole), 51cb, Ekaterina Mikhailova 61tr, Mirek1967 43cra, Mirkorosenau 38cra, Mouse Family Mouse Family 25br, Natalya Aksenova / Natalyaa 31cb (duck), Pavel Naumov 53bc, Sivakorn Nayanetra 31bl, Neirfy 30clb, 31br, Yin Jian Ng 35bc, Nivanova250788 62cl (bigfoot), Duncan Noakes 16cr, Nostradamus252 34br, Rungroj Nuiman 43br (Antx2), Nyker 16c, Nylakatara2013 42ca, Veronika Oliinyk 1clb, Olga Itina / Olikit 35tr, Onyxprj 63cr (elf), Eline Oostingh 18bc, Ornitolog 12ca, Oskanov 43tc, Oxilixo 31bc, Paleka 50c, Kevin Panizza / Kpanizza 9bl, Juan Bautista Ruiz Páramo 43ca, Parfentevamaya 21cla, Dmytro Parkheta 43cl, Gueret Pascale 59crb, Prosun Paul 11tl, Maksim Pauliukevich 34cla (dust), 35ca (dust), Kostya Pazyuk 25c (rabbit), Peerapong Peattayakul 24ca, Martin Pelanek 43cr, Azahara Perez 61clb, Stefan Hermans / Perrush 49cla, Photoclarity 25bl, Photoeuphoria 41cr, Pimmimemom 47clb (Monarch), Peter Leahy / Pipehorse 59cra, Elena Podolnaya 42crb, Stu Porter 12bl, Alexander Potapov 31cb, Grobler Du Preez 43tl, Ondřej Prosický 19ca, 38cla, 40cra, 59tl, Pytyczech 1tc, 56br, Rogerio Queiroz 43tc (nest), Alexander Raths 20cr, Mohd Zaidi Abdul Razak 39tc, Ian Redding 47tl, Rhallam 30cra, Francesco Ricciardi 6br, 39c, Dan Rieck 59cb, Rikke68 3cb, 34tr, Rinus Baak / Rinusbaak 32ca, 38ca (Macaw), Eurico Rodrigues 39bl, Craig Russell 43crb, Kaewmanee Saekang 7clb, Samum 49fcla, Sarah2 8crb, Seaonweb 38br, Anna Sedneva / Sedneva 53b, Inha Semiankova 24ca (pond), Yury Shirokov 31bc (Cat), Pavel Shlykov 27clb (Russell), Andrei Shupilo 8cla, 8cl, Slowmotiongli 7crb, 13cb, 19tl, 35ca, 47cb, 51tl, 55cla, Simone Gatterwe / Smgirly 30bl, Smileus 22br, Michael G Smith 6cla, Olya Solodenko 24cra (bed), David Steele 43ftl, Andreas Steidlinger 43bl, Studio 37 / Dreamstock 38cr, Stu Porter / Stuporter 29tl, Kedsirin Suthamsakul 51crb, Tartilastock 34clb (trees), Taviphoto 12bc (duck), Thawats 34tr (monarch), 34cra (flutter), Charoenchai Tothaisong 55cl, Trinhhuytho 35bl, Troichenko 11c, Sergey Uryadnikov 17cr, 19bc, Vac 17br, Veleknez 31c, Venturebeyond 47tc (Squid), Verastuchelova 24br, 49cl, Gale Verhague 10br, Vasiliy Vishnevskiy 49cla (Rat), Viter8 30br, Vaclav Vitovec 10cla, Vladvitek 12bc, 21tc, 39cl, Yehor Vlasenko 2br, 62-63bc, Tomislav Vucic 43c, Wafuefotodesign 49cb, Wahyudinfirman 52bc, Gary Webber 40tr (texture), Welcomia 59bc, Ashley Whitworth 7cra, Buddee Wiangngorn 9cla, 9bl (sand), 37b, Apisit Wilaijit 41clb, Wildlife World 13cb (sparrow), Jan Martin Will 59cl, William Wise 50cb, Marcin Wojciechowski 1cb, Wollertz 34clb, Jinfeng Zhang 11cla, Katerina Zmachynska 63cla (ship), Rudmer Zwerver 15crb, 17tc, 34clb (Kingfisher), 53cla; **Fotolia:** anankkml 55ca, giuliano2022 6cl, Eric Isselee 44br, Sergey Khachatryan 49br, Andrey Eremin / mbongo 47tl (Waves), xstockerx 64tr, Stefan Zeitz / Lux 12cl; **Getty Images:** Collection Mix: Subjects / Michael Nolan 37tr, Brandon Tabiolo / Design Pics 6cb, Sylke Rohrlach / EyeEm 37cb (slug), imageBROKER / Reinhold Schrank 43bc, Moment / Stan Tekiela Author / Naturalist / Wildlife Photographer 36cra, Moment / Tambako the Jaguar 48ca, Stone / Michael Duva 35cr, Westend61 21br; **Getty Images / iStock:** 2630ben 17tl, alkir 22cra, Andyworks 28bc, Antagain 1tl, 31crb (bees), Aunt_Spray 60tr, Kristian Baensch 44cla, BionicPanda 63br, Boyshots 44bl (wombat), Vicky_Chauhan 51br, CoreyFord 44ca, defun 22cb, dennisvdw 19tr, sserg_dibrova 36-37bc, DigitalVision Vectors / exxorian 62br, E+ / 4FR 53tr, E+ / Kativ 47br, E+ / KenCanning 16cla, 44clb, E+ / Raycat 56crb, E+ / vusta 23crb, Entwicklungsknecht 35c, feedough 52cla, Liliya Filakhtova 10cl (beetle), Flexire 39crb, FrankRamspott 50ca, girlfrommars 62cb, GlobalP 31tl, 31ca, 41cla, Grisha459 55cr, Taisiia Iaremchuk 63tl, Kaphoto 30cr, KeithSzafranski 44tr, Piotr Krzeslak 30bc, Lanaclipart 62cla, leonello 61cra, lillybell 44bc, Lidiia Lykova 44cb, micro_photo 57cl, mtruchon 30ca, Hachio Nora 62cl, Placebo365 43cla, proxyminder 56l, reptiles4all 22cb (terrapin), Chelsea Sampson 38clb, Victoria Shapkina 63bc, Sieboldianus 55cb, 55crb, Christophe Sirabella 20clb, stanley45 22cla, studiocasper 48clb, SurfUpVector 62crb, Tazzy1 29bc, undefined 18crb, vendys 35fcrb; **naturepl.com:** Philip Dalton 37cla, Georgette Douwma 36bc, 37br, Tim Laman 37ca, Thomas Marent 28cra, 36c, Alex Mustard 36br (limpet), 37bc, MYN / Brett Lewis 15cb, Piotr Naskrecki 37tl, Nature Production 37cra, Gary Bell / Oceanwide 32bc, Pete Oxford 36br, Morley Read 15br, Andy Sands 32tr, David Shale 36bc (sponge), 37cb, Nick Upton 42c (holt), Doug Wechsler 14ca, Rod Williams 28clb, Solvin Zankl 32bl; **Science Photo Library:** Mauricio Anton 60cla, Nicholas Bergkessel, Jr. 35cla, British Antarctic Survey 36cb, Robert Chase 61cb (seal), Dennis Kunkel Microscopy 6cra, K Jayaram 37ca (frog), Andrew J. Martinez 50br, Cordelia Molloy 29br, Nicholas Smythe 36ca, Roman Uchytel 60cl, 61cl; **Shutterstock.com:** Kurit afshen 55c, Dray van Beeck 39cr, Billion Photos 24cb, Jude Black 10tr, BRO.vector 63bc (harpy), Cingular 6cr, Jesus Cobaleda 39bl (parrotfish), 50-51ca, delcarmat 62bc, Dirk Ercken 1cra, Erni 16cl, Gerald Robert Fischer 49crb, Gallinago_media 48c, Anton Kozyrev 10cra, MongPro 48cra, Mr.Photomato 42cra, NickEvansKZN 23tl, panpilai paipa 24cla (fish), RealityImages 41crb, Porco_Rosso 59br, Sarah2 11cl, sonelle.vdm 43tc (Weaver), I Wayan Sumatika 60br, Pavaphon Supanantananont 38br (basslet), vkilikov 20cl, Wirestock Creators 11bl, Michiel de Wit 14cr, chonlasub woravichan 38bc, xpixel 25tr, Milan Zygmunt 54c; **SuperStock:** Biosphoto / Gregory Guida 46cb, Science Photo Library 6c; **Didier Descouens, Museum of Toulouse:** 46cr

Cover images: *Front:* **123RF.com:** Thawat Tanhai (Kingfisher), Pavlo Vakhrushev / vapi (jellyfish); **Dorling Kindersley:** Peter Chadwick / Natural History Museum, London (skull), Gary Ombler / Royal Botanic Gardens, Kew (leaves); **Dreamstime.com:** Isselee (prairie), Jianghongyan (clam), Nyker (Llama), Olga Itina / Olikit (horse), Palex66 (insect), Kevin Panizza / Kpanizza (Sponge), Pytyczech (Morpho), Rinus Baak / Rinusbaak (Bat), Sarah2 (tick), Yobro10 (Elephant), Rudmer Zwerver (Mouse); **Fotolia:** xstockerx (GPig); Front and Back: **123RF.com:** Eric Isselee / isselee (Koala); **Dorling Kindersley:** Jerry Young (Gramma), (tetras), (Spider); **Dreamstime.com:** Veronika Oliinyk (footmarks), Korn Vitthayanukarun (texture), Linda Bucklin (Whale), Farinoza (bushbaby), Irisangel (feather), Isselee (Clownfish), (Cockatoo), Johannesk (Moorish), Dirk Ercken / kikkerdirk (frogx2), Irina Kozhemyakina (opossum), Brian Kushner (Eagle), Alexander Potapov (duck), Rikke68 (Buzzard), Studio 37 / Dreamstock (Redfish), Marcin Wojciechowski (Reindeer); **Fotolia:** Stefan Zeitz / Lux (puffin); **Getty Images / iStock:** anankkml (puma), Antagain (parrot), chris2766 (hen), GlobalP (Axolotl), (fox), Kaphoto (Ant); **Shutterstock.com:** Dirk Ercken (frog); *Back:* **123RF.com:** Eric Isselee / isselee (Koala); **Dorling Kindersley:** Frank Greenaway / Natural History Museum, London (Moth), Robert Royse (Willow); **Dreamstime.com:** Annaav (Fennec), Ben (Peacock), Brad Calkins / Bradcalkins (snail), Linda Bucklin (Whale), Farinoza (bushbaby), Irisangel (feather), Isselee (spider), (Clownfish), (Cockatoo), (Tortoise), Rinus Baak / Rinusbaak (Macaw); Spine: **Dreamstime.com:** Johannesk (Moorish); **Shutterstock.com:** Dirk Ercken (frog)

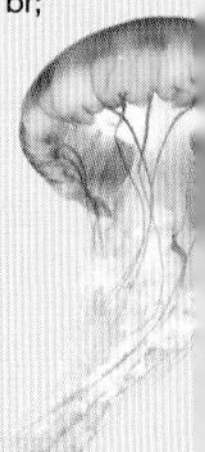